HARCOURT SCIENCE

VIRGINIA SOL SUPPORT FOR STUDENTS

GRADE 3

Harcourt

Orlando   Austin   Chicago   New York   Toronto   London   San Diego

Visit *The Learning Site!*
www.harcourtschool.com

Reviewers

Sharon Bowers
Kemps Landing Magnet School
Virginia Beach, Virginia

Brenda Dorman
Coleman Place Elementary
Norfolk, Virginia

Coleen Matthews
Coleman Place Elementary
Norfolk, Virginia

Diane C. Tomlinson, Ed. S.
Virginia Science Content Specialist/Codirector
Coalfield Rural Systemic Initiative
Lebanon, Virginia

Contents

LESSON 1

Months and Seasons

Find Out

- about changing seasons
- how plants change with the seasons

Vocabulary	
cycle	winter solstice
spring equinox	summer solstice
fall equinox	life cycle

Changing Seasons

Many wildflowers bloom in the spring. Many trees lose their leaves in the fall. Seasons affect wildflowers, trees, and other living things. Seasons occur in a **cycle**, a pattern that repeats itself. Winter, spring, summer, and fall follow a cycle year after year.

In Virginia each season lasts about three months. Winter usually begins on December 22. This day has the fewest hours of daylight. It is called the **winter solstice** (SOHL•stis). After the winter solstice, the number of daylight hours increases each day. On about March 21, the day has exactly 12 hours of daylight and 12 hours of darkness. This day is called the **spring equinox** (EE•quih•nahks). It is the beginning of spring.

Just as there is a winter solstice, there is a summer solstice on about June 21. The **summer solstice** has the greatest number of daylight hours of any day in the year. It is the beginning of summer.

After the summer solstice, the number of daylight hours begins to decrease. On the day in September that there are exactly 12 hours of daylight and 12 hours of darkness, the fall season begins. This is the **fall equinox**. The date is September 22 or 23. Three months later, it is winter again.

✔ **1** What is the meaning of the term *equinox*?

Virginia and Seasons

Because each season in Virginia is different, you can see many changes in animals and plants. Spring in Virginia is warm, and the number of daylight hours increases. As the weather gets warmer, plants begin to grow. For example, dogwood trees begin to have leaves, as well as buds that will open into flowers.

During the summer the weather gets hot, and there are many hours of daylight. Plants normally get a lot of light and rain, so they become very healthy. Dogwood trees are full of green leaves.

In the fall there are fewer hours of daylight and the weather gets cooler. Leaves slowly change in color and then fall to the ground.

Virginia winters are very cold. Most plants cannot grow. Many trees, including dogwood trees, have bare branches because the leaves have fallen off.

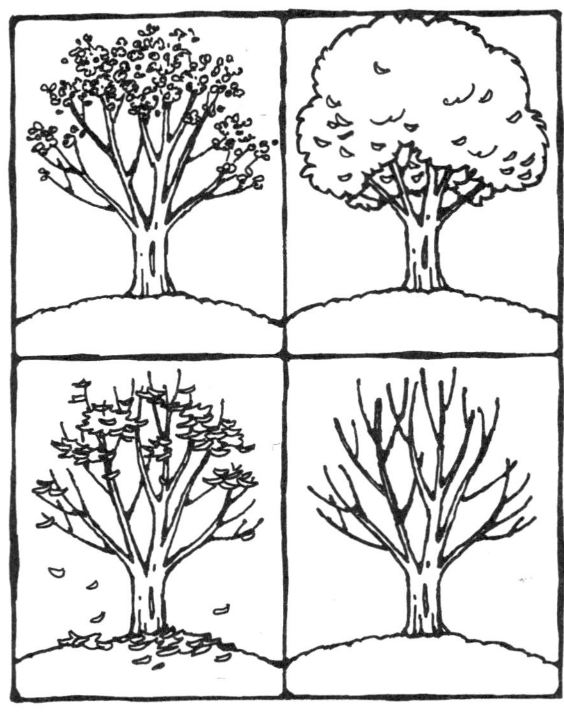

Trees change with the four seasons.

✔ **2** How do the changing seasons affect the dogwood tree?

Plant Cycles and Seasons

Plants have cycles that are connected to the seasons. All the changes that a plant goes through—from a seed, to its full growth, to its death—make up the plant's **life cycle**. Some plants complete their whole life cycle in one year. Other plants live for many years.

Some plants are used for food, such as beans and peas. Think about soup beans or baked beans. The beans are seeds. In the spring, when temperatures get warmer and seeds get water, the seeds germinate. Small plants begin to sprout from the seeds. Soon seedlings form. During the spring or summer, the plants are fully grown and flowers form. In the summer and fall, the plants develop fruits, which hold the new seeds. Then the bean plants die.

A bean plant goes through its whole life cycle in one year. When beans are planted in the spring, the life cycle starts again.

This bean plant begins its life cycle as a seed. Once the seed sprouts, a bean plant begins to grow.

Other plants, like the dogwood tree, live for many years. Each year they produce flowers and seeds. Some seeds will grow into new trees. When an old dogwood tree dies, its life cycle is complete.

✔ ❸ How does the life cycle of most plants begin?

Activity

Directions The pictures show part of the life cycle of an apple tree.
Put the pictures in the correct order. Number them from 1 to 7.

**Flowers form on
the tree.**

**The apple tree
grows.**

**Fruits begin to form
on the tree.**

**The seeds
germinate.**

**An apple tree sprouts
above the ground.**

**The apple seeds are
planted.**

**Seeds form inside
the fruits.**

LESSON 2

Animal Life Cycles

Find Out

- about the life cycles of chickens, monarch butterflies, and mammals

<table>
<tr><td>Vocabulary</td></tr>
<tr><td>hatches   larva   pupa   mammals</td></tr>
</table>

Animals Have Life Cycles

When you look at a picture of yourself as a baby, you can see that you were very different. You have grown since you were a baby. This growth is part of your life cycle. The life cycle of an animal is the order of stages, or periods, in its growth. Most animals begin their life cycle as an egg. Even dogs, cats, and humans start out as egg cells.

✔ **1** How do most animals begin their life cycle?

The Life Cycle of a Chicken

Adult female chickens are like other birds. They lay eggs with a hard shell, or outer covering. The chick growing inside an egg is protected by the shell. When the chick inside grows big enough, it **hatches**, or breaks through the eggshell. The chick grows into an adult. If it is a female, it can then lay eggs, and the cycle begins again.

The chick breaks out of its egg. It no longer needs its shell for protection. It soon grows into an adult chicken.

✔ **2** What is a function of an eggshell?

The Life Cycle of a Butterfly

A monarch butterfly goes through four different stages during its life cycle. First, an adult female butterfly lays eggs. A butterfly egg then hatches into a caterpillar, also called a **larva**. After being in this larva stage, the caterpillar makes a hard covering around itself, called a **pupa**. The pupa changes and then a butterfly comes out. The adult female lays eggs and the cycle begins again.

Butterflies go through big changes in their life cycle. When a butterfly begins its life cycle, it doesn't look like an adult butterfly at all.

✔ **3** What are the four stages in the life cycle of a monarch butterfly?

The Life Cycle of a Mammal

Animals that have fur or hair are called **mammals**. Humans are mammals, and so are dogs, cats, and cows. Most mammals form and grow inside the mother's body. The young are born live from their mother. After birth the mother produces milk to feed the young. Young mammals grow into adults that produce more young.

✔ **4** How do mammals feed their newborn young?

Review

5 What differences are there between the life cycles of birds and mammals?

Activity

Choose an animal that lives in Virginia, such as the white-tailed deer. Do research so you can draw and label the stages in the animal's life cycle. Cut out each stage, and place your drawings in a bag along with the other students' drawings. Take turns pulling a picture out of the bag. Work together to put the drawings in order. Share your completed life cycles with the class.

LESSON 3

Diseases That Harm Plants

Find Out

- about diseases that have killed American trees

> **Vocabulary**
>
> fungus (plural, fungi)
> Dutch elm disease   chestnut blight

Plant Diseases

What do you think would happen if a plant got sick? Its leaves might wilt and it might die. Plants can get sick, just as people can. In both people and plants, many diseases are caused by germs that can be spread from one individual to another. Many plant diseases are caused by fungi (FUHN•jy). A **fungus** (FUHNG•guhs) is an organism that looks a little like a plant but does not make its own food. It feeds on other organisms, and if it is living on a plant, it can weaken or kill the plant.

✔ ❶ What kind of germ causes many kinds of plant disease?

Dutch Elm Disease and Chestnut Blight

American elm trees once lined many of the town streets in Virginia. There also used to be many American chestnut trees. Now most of those trees are gone. Many of those trees have suffered from diseases that came from other countries.

Elm trees were taller than most houses and provided shade on hot summer days. Today, because of a deadly fungus that causes **Dutch elm disease**, most of those trees have died. This fungus arrived on a shipment of wood from Holland in 1930. When a tree has Dutch elm disease, the tubes that carry water up the trees become clogged. The fungus can kill an elm tree in only three weeks.

America's eastern forests were once filled with American chestnut trees. Around 1900 a fungus entered the United States from Asia. Wind, rain, birds, and other animals spread it through the forests.

The fungus entered the trees through openings in the bark, and then it multiplied. It cut off the trees' supply of water and nutrients. The disease it caused, **chestnut blight**, usually kills a tree in just one year. By 1940 3 billion American chestnut trees had died. Today the blight fungus is still around. Some smaller American chestnut trees can be found. However, they are still being attacked by the fungus.

When trees like this chestnut get chestnut blight, they may die very quickly.

✔ ❷ What are the causes of disease for Virginia's elms and chestnuts?

Activity

Directions **Choose the elm, the chestnut, or another type of Virginia tree, and give details about its disease. Tell how the tree and the ecosystem change once the tree gets the disease. Complete a chart like the one below.**

Details About an American Tree in Danger	
Type of tree	
Cause of disease	
How disease affects tree	
Other plants affected by this disease	

LESSON 4

Alternative Sources of Energy

Find Out

- why trees are an important renewable resource
- about nuclear energy
- about renewable sources of energy

Vocabulary
timber   nuclear energy

Trees as a Resource

Energy is important for everything you do. Many communities get energy from electrical plants that use coal or oil to produce electricity. But there are other ways to produce electricity or to heat a house. For example, wood is used as a fuel in many parts of the world. People use the wood from trees to heat their homes and to make fires for cooking food.

In the United States, however, trees are used mainly for timber. **Timber** is wood used in building homes, offices, and other buildings. It is also used to make furniture. Another important use of trees is for making paper. Books, newspapers, and magazines are all made from wood. In addition, some fruits and spices come from trees.

Trees are important renewable resources. Think about what happens when trees are cut down. As long as some trees are left behind, new trees will grow. However, sometimes forests are completely cut down to make room for roads or buildings. What will happen then to the supply of wood? Animals that once lived in the forests must find new homes. Where will the animals go?

What would your own life be like without trees? There would be no pencils, no paper, no warm nights by a wood-burning fireplace.

✔ **1** What are two ways in which wood is used?

Many of these items are made by using things that come from trees. Identify all the items that are made of wood. Think about other items around your school and your home. How many things are made of wood?

Nuclear Energy

Another source of energy is nuclear energy. **Nuclear energy** is produced by splitting very small particles of matter called atoms. The place where nuclear energy is produced is called a nuclear reactor. Nuclear reactors produce huge amounts of energy. The energy is released as heat, which is then used to produce electricity.

However, nuclear energy has some problems. One problem is that it produces dangerous waste. Scientists have to find a way to store the waste so it won't hurt people or the environment. Nuclear energy is also more expensive than people once thought it would be. This is because nuclear reactors must be carefully built and maintained to make them safe.

✔ ❷ What is the problem with waste from producing nuclear energy?

Electricity from Wind and Water

Wind is another source of energy. Wind can be used as an energy source in some places, with the help of huge windmills. These windmills, sometimes called wind turbines, use the energy of the wind to produce electricity. Wind turbines may have blades 25 meters (82 ft) long! Wind turbines are usually set up in large groups. These are called wind farms. They are in places where the wind blows hard almost every day.

Wind farms have windmills that convert the energy from the wind into electricity.

Dams are often built on rivers. The water behind a dam forms a large lake. Some of the water from the lake flows through the dam. The energy in the flowing water is used to make electricity. In some parts of the United States, the energy of flowing water provides electricity for whole towns.

As the water from this lake flows to the stream below, it can produce electrical energy.

✔ ③ What kind of energy can we get from wind and water?

Activity

Directions For each type of resource, list why it is good to use and why it might not be good to use.

	Why It Is Good to Use	Why It Might Not Be Good to Use
Wood		
Nuclear Energy		
Wind		
Water		

LESSON 5

Resource Renewal

Find Out

- the difference between endangered and extinct species
- the importance of resource renewal programs

<table>
<tr><td>Vocabulary</td></tr>
<tr><td>endangered species
extinct
resource renewal</td></tr>
</table>

Endangered Species

The smooth coneflower once grew in many parts of Virginia. Now it is hard to find. This plant is an **endangered species,** a species in danger of dying out completely. If no one protects it, the plant will be **extinct**.

Animals can be endangered, too. Some woodpeckers, bald eagles, and many types of whales are in danger of becoming extinct.

 **1** What is an endangered species?

Saving Endangered Species

Resource renewal is a type of conservation in which people protect and help endangered species. The seeds of endangered plants can be collected and planted indoors. The new plants are then planted outside in their natural habitats. Scientists can capture endangered animals and breed them. Then they release them in the wild.

 **2** What is resource renewal?

Endangered species need protection. The smooth coneflower and bald eagle both need help.

Activity

Directions The map shows the number of endangered species per state across the United States. Use the map to answer the questions.

1 Name four states with 10 or fewer endangered species.

____

2 Name the two states that have the most endangered species.

____

3 How many endangered species are there in Virginia?

____

4 In which state is a resource renewal program likely to be more helpful, Maine or California? Why?

____

____

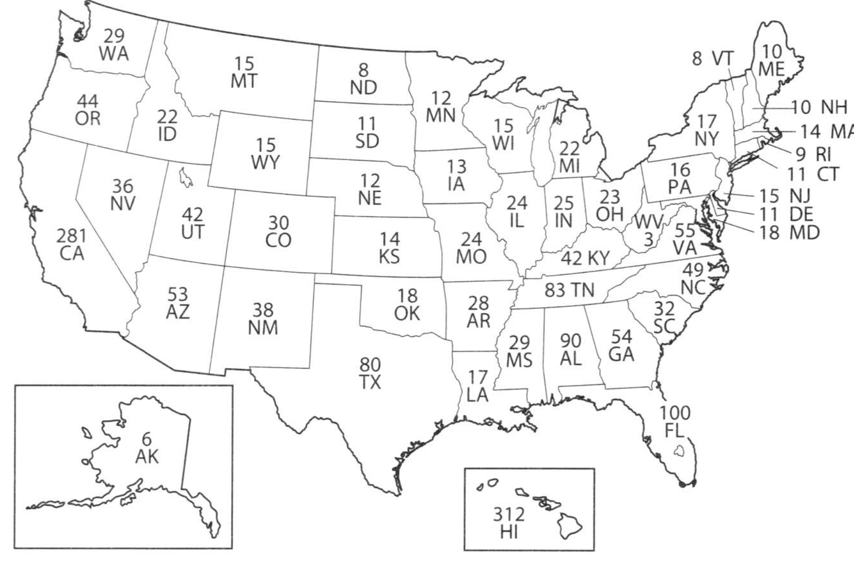

The Number of Endangered Species per State

LESSON 6

Time and Tides

Find Out

- what causes the tides
- about high and low tides
- what affects the tides

Vocabulary
tides   gravity   high tide
low tide   tide tables

The Rise and Fall of the Oceans

Suppose you are at the beach. You build a sand castle close to the water. Then you go swimming. Afterward you have lunch and play ball with your friends. By then the water has moved up the beach. The waves have eaten away at your castle. It is almost gone.

Ocean water rises and falls in regular cycles called **tides**. Your sand castle has been washed away by the rising tide.

✔ **1** What are tides?

What Causes Tides?

Gravity is a pull of all objects toward each other. The force of gravity causes tides in Earth's oceans.

Both the moon and the sun pull on Earth. The sun is bigger, but the moon is much closer to Earth. That is why the pull of the moon's gravity is the main cause of the tides.

The moon's gravity pulls on the ocean water on Earth, and the ocean water forms bulges. The changing bulges are what make the tides.

✔ **2** What is the result of the force of gravity between two objects?

High and Low Tides

Where is it high tide or low tide? A **high tide** happens at a point on Earth where the water bulges. A **low tide** happens at a point on Earth between the bulges.

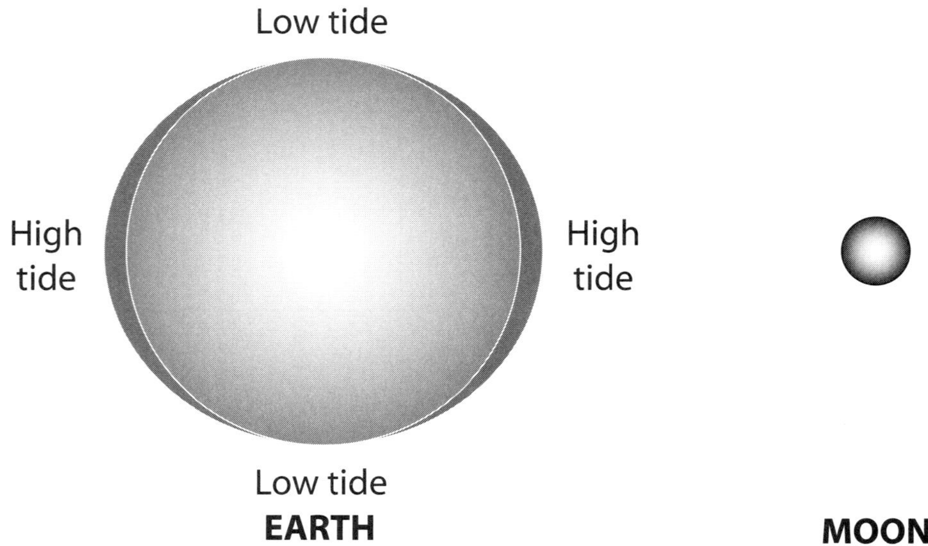

High tides occur on the sides of Earth that are facing toward the moon and facing away from the moon. Low tides occur in between.

Most coasts have one or two high tides and one or two low tides each day. If you stayed at Virginia Beach for a little more than 24 hours, you could see the tides rise and fall two times. In most places on Earth, it takes about 6 hours to go from high tide to low tide. It then takes another 6 hours for the tide to become high again.

Of course, tides in different places don't all happen at the same time. When it is high tide in Virginia, it is closer to low tide across the ocean in Africa. Also, tide heights are not the same everywhere. Even in places close to each other, the tides might be different in height.

✔ ❸ At any one time, where on Earth does high tide occur?

Tides Differ Greatly

Things besides the moon and the sun affect the tides. These things include the shape of the coastline, the shape of the sea floor, and the depth of the water.

In some places you can hardly tell the difference between high tide and low tide. Out in the open ocean, for example, it is hard to notice the tide changing. At some places on Earth, however, the difference between high and low tide is very large.

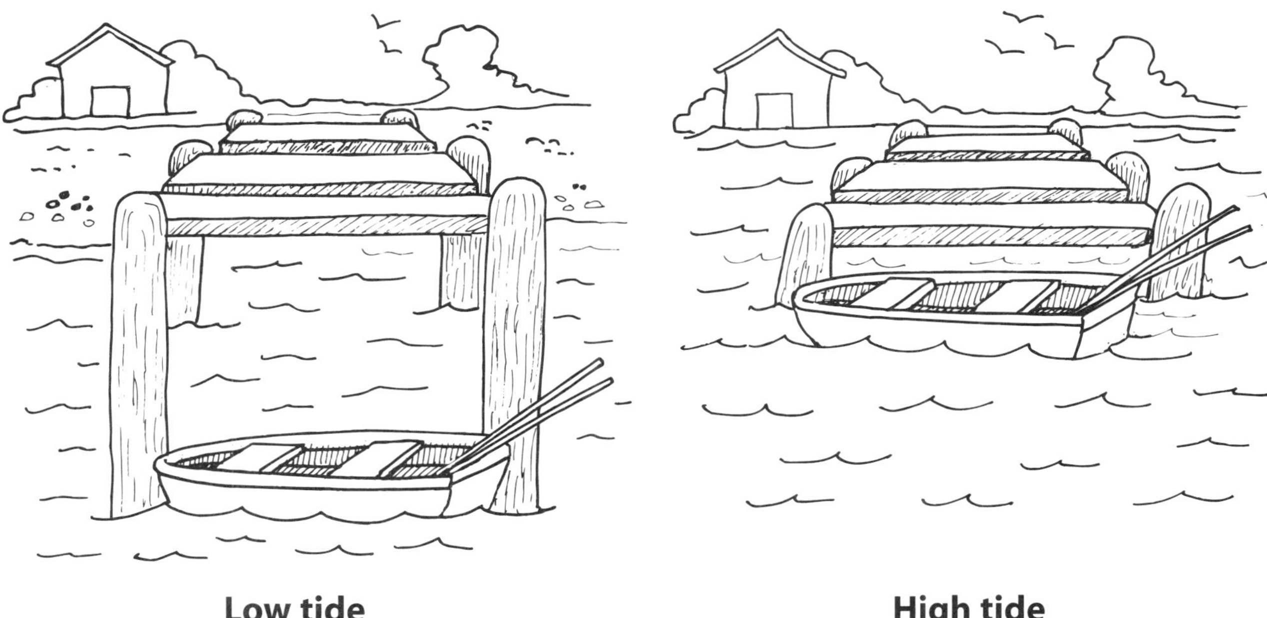

Low tide   **High tide**

The tides at the Bay of Fundy, in Nova Scotia, can make the water rise as much as 15 meters (about 50 ft)!

✔ **4** Name two things, besides the moon and the sun, that can affect the tides.

Activity

Tide Tables

The tides can be predicted. Newspapers print the times of high and low tides for each day. These tide predictions are placed in **tide tables**. Tides are affected by many things. In this activity you will see how the heights of tides are related to the phases of the moon.

Directions Each day for two weeks, record the information from a tide table. Write the information in the chart below. Tide tables can be found on the weather page of many newspapers or on the Internet. Mark high tides with an *H* and low tides with an *L*. What do you notice about the phases of the moon and the heights of the tides?

Location of the tide station: _________________________________

Date	Height of Tide 1	Height of Tide 2	Height of Tide 3	Height of Tide 4	Moon Phase

© Harcourt

LESSON 7

Measuring Time

Activity Purpose How long does it take you to get to school? How long does it take you to get from your classroom to the cafeteria? Knowing how long it takes for something to happen is often helpful. How long do you think it would take for an ice cube to melt? Would it always take the same amount of time? What things might affect the melting? Investigate to see how **measuring** time can give you information about ice cubes melting.

Materials

- 2 paper towels
- white foam cup
- white plastic plate
- 2 ice cubes of the same size
- clock or watch

Activity Procedure

1 Place the cup on one paper towel. Put the plate on another paper towel. Put one ice cube in the cup. Put the second ice cube on the plate.

2 **Form a hypothesis** about how long it will take the two ice cubes to melt.

3 Make a table like the one shown to **record** your observations. Record the starting time to the nearest minute.

4 **Observe** the ice cubes every 5 minutes.

5 When each cube is totally melted, **record** the time.

Observations of Melting Ice Cubes		
	Starting Time	**Finish Time**
Ice Cube in Cup		
Ice Cube on Plate		

Draw Conclusions

1 How long did it take each ice cube to melt? _______________________

2 How did the melting time of the two ice cubes compare? _______________

____

3 What do you think caused the melting time of the two ice cubes to be the same or different? _______________________________________

____

4 **Scientists at Work**   Scientists test only one **variable** at a time. What variable did you test in your experiment? _______________________

Investigate Further   Plan and conduct an experiment to test this hypothesis: The temperature affects how fast ice melts.

LESSON 8

Make a Compound Machine

Activity Purpose Have you ever tried to lift or move something that was very heavy? People are faced with this problem all the time. How can we move heavy things easily, without hurting ourselves? Machines make work easier to do. In this investigation, you will make a compound machine to help you move an object. A compound machine is made up of two or more simple machines.

Materials

- small pulley
- toy car
- masking tape
- string, 1 m
- spring scale, 10-N
- meterstick or meter tape measure
- ramp (small board, 10 cm wide × 30–40 cm long)
- books

Activity Procedure

Part A

1. Tie the string to the car. Make a loop in the end of the string and hook the spring scale through it.

2. Make a ramp with the board and some books. It should be about 10 cm high.

3. Put the car at the bottom of the ramp. Pull the spring scale to move the car up the ramp. **Record** the number on the spring scale.

4. Repeat Step 3 twice. Remember to **record** your data.

Part B

5 Untie the string from the car. Attach the pulley to the car. Put the string through the pulley. The spring scale should still be attached to the end of the string.

6 Hold the free end of the string against the top of the ramp.

7 Repeat Steps 3–4.

Draw Conclusions

1 Were the spring scale readings higher in Part A or Part B?

____

2 Which machine required less effort to lift the car?

____

3 What are the machines that make up the compound machine in Part B?

____

4 **Scientists at Work** Scientists must **measure** carefully to gather good data. What did you use to measure forces in this activity?

____

Investigate Further Suppose you were trying to lift a mass much heavier than the car in this activity. Describe how you would change the materials used in this activity to lift the mass.

LESSON 8

Simple and Compound Machines

Find Out

- about the six kinds of simple machines
- what a compound machine is

Vocabulary	
simple machine	inclined plane
wedge	lever fulcrum
wheel and axle	pulley screw
compound machine	

Simple Machines

A **simple machine** is a tool that helps people do work. A simple machine makes a job easier to do. It doesn't change the amount of work needed to do the job. In your *Harcourt Science* textbook you studied work on pages F74–75. Remember that work happens when a force acts on an object over a distance.

 **1** What is a simple machine?

Inclined Plane

You have probably seen a wheelchair ramp. A ramp is a simple machine called an inclined plane. An **inclined plane** is a flat surface set at an angle to another surface. It is easier to lift something with an inclined plane because less force is needed to lift the object. However, the same amount of work is done because the force is used over a longer distance.

2 What is an inclined plane?

The inclined plane on a mover's truck makes it easier to lift heavy furniture.

Wedge

When you cut food with a knife, you are using a wedge. A **wedge** is two inclined planes stuck back to back. It is thick at one end. The other end is a thin edge. A doorstop is another example of a wedge.

A long, thin wedge makes a job easier than a short, thick wedge. With the thin wedge less force is needed to do the same amount of work.

A wedge can be moved. For example, an axe head is one kind of wedge. It is attached to a handle. A person splitting wood uses a force to move the handle by swinging it. The handle gives a force to the thick end of the wedge. This force pushes the wedge down into the log. The wedge puts a force on the log and splits it in two.

A saw is made up of many tiny wedges in a row. Each little wedge splits off a small bit of wood.

 **3** Give three examples of wedges.

The head of an axe is a wedge. This wedge makes splitting wood an easier job.

Screw

A **screw** is like an inclined plane wrapped around a pencil. The spiral inclined plane makes up the thread of the screw. You can use a screwdriver to twist a screw into a piece of wood. As the screw is turned, the inclined plane on the screw pulls it into the wood.

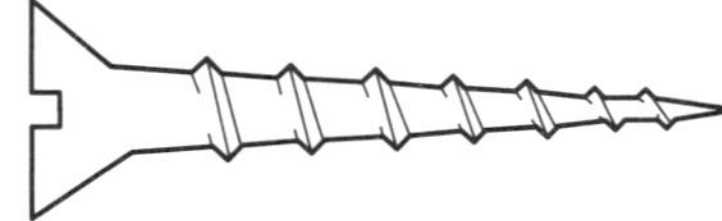

Simple machines come in all sizes. They can be small, like this screw.

What are some other tools that make use of screws? Think of the things you screw on and off. A jar lid is one example of a screw.

 **4** How is a screw like an inclined plane?

Lever

A **lever** is a bar that moves on or around a fixed point called a **fulcrum**. A seesaw is one kind of lever. On a seesaw the bar is the board you sit on. As the board moves on the fulcrum, you go up and down.

If you have ridden on a seesaw, you have used a lever.

A paint-can opener is another lever. The opener rests on the edge of the can. The edge is the fulcrum. The tip of the can opener is under the lid. You push down on the handle of the opener. The opener moves on the fulcrum and the tip pushes upward and opens the can. You have to push the handle a longer distance than the lid moves. But it takes less force than pulling up the lid would take. The work is easier.

Levers are all around you. Examples include baseball bats, hammers, fishing poles, and pliers.

✔ **⑤** What is a lever?

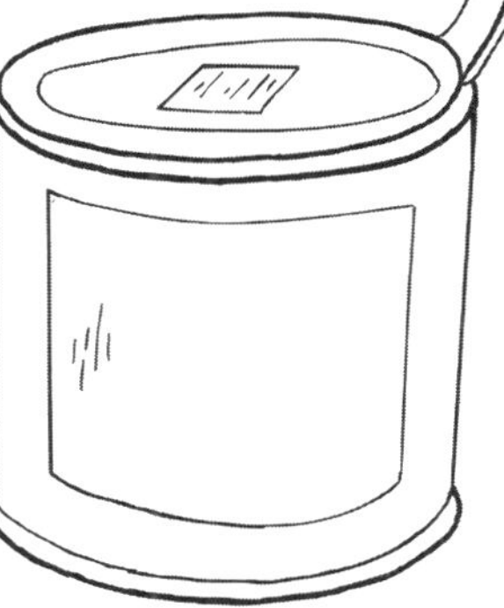

If you have used a paint-can opener, you have used a lever.

Wheel and Axle

Imagine twisting a screw into wood with your fingers. It would be hard, wouldn't it? A screwdriver makes the work much easier. A screwdriver is an example of a wheel and axle. A **wheel and axle** is made up of two circular objects that are fastened together. The larger object is the wheel. The smaller object is the axle. The wheel and the axle move together.

In a wheel and axle, less force is needed to move the wheel than to move the axle. When the wheel makes one full circle, it goes a lot farther than the axle. In a screwdriver, the handle is the wheel. The metal shaft is the axle.

A doorknob is also a wheel and axle. The knob is the wheel. The metal shaft that goes into the door is the axle. When someone pedals a unicycle, the cranks that are turned act like a wheel around an axle.

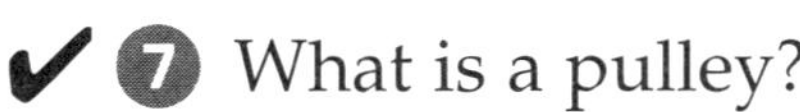

Imagine riding on just one wheel and axle!

✔ **6** Which part of a screwdriver is the wheel?

Pulley

A **pulley** is a rope over a wheel. It changes the direction of the force you use. When you pull down on one end of the rope, the object attached to the other end moves up. Some pulleys use a chain or a steel cable instead of a rope.

A pulley is used to raise and lower a flag on a flagpole. A pulley is also used to raise and lower a shade on a window. Neither of these pulleys changes the amount of force needed.

✔ **7** What is a pulley?

Pulleys help raise and lower objects by changing the direction of a force.

© Harcourt

Compound Machines

Many tools and machines are made up of more than one simple machine. Two or more simple machines put together form a **compound machine**. Look at the diagram on page F80 in your *Harcourt Science* textbook. The pair of scissors is made up of two levers and two wedges. Each blade of the scissors is a wedge. A bolt near the center of the two blades holds them together. The bolt is a type of screw. Each wedge moves around the bolt. This turns the wedges into levers. When force is used on the levers, the narrow edges of the wedges meet and can cut through paper or fabric. So, a pair of scissors is a compound machine made up of five simple machines.

A staple remover is another kind of compound machine. It is also made up of two levers. One end of each lever has wedges for taking out the staples.

A tape dispenser is also a compound machine. The roll of tape is on a wheel and axle. When you pull on the tape, the wheel and axle moves to allow more tape to come out. Then you cut the tape on a cutting edge made up of many small wedges.

Because it is made up of simple machines, this tape dispenser is a compound machine.

A wheelbarrow is another kind of compound machine. What simple machines make it up?

✔ **8** What is a compound machine?

Summary

Simple machines make it easier to do work. They reduce the amount of force needed or change the direction of a force. There are six kinds of simple machines. They are the inclined plane, wedge, screw, lever, wheel and axle, and pulley. Compound machines are made up of two or more simple machines.

Review

9 Give an example of each kind of simple machine.

10 How does a pulley help with lifting something?

11 You have a choice between a short inclined plane and a long one. Which would you choose? Why?

Activity

A pencil sharpener is a compound machine. It is made up of a wheel and axle and wedges. Label each of the simple machine parts in the following diagram: